Andréa de Freitas Souza
Raymison Cardoso
Danylo de Araujo Viana

Economic and environmental feasibility study

Andréa de Freitas Souza
Raymison Cardoso
Danylo de Araujo Viana

Economic and environmental feasibility study

For the use of scrap tyres in the asphalt paving
of a federal highway and an avenue in Natal/RN

ScienciaScripts

Imprint

Any brand names and product names mentioned in this book are subject to trademark, brand or patent protection and are trademarks or registered trademarks of their respective holders. The use of brand names, product names, common names, trade names, product descriptions etc. even without a particular marking in this work is in no way to be construed to mean that such names may be regarded as unrestricted in respect of trademark and brand protection legislation and could thus be used by anyone.

Cover image: www.ingimage.com

This book is a translation from the original published under ISBN 978-613-9-66121-3.

Publisher:
Sciencia Scripts
is a trademark of
Dodo Books Indian Ocean Ltd. and OmniScriptum S.R.L publishing group

120 High Road, East Finchley, London, N2 9ED, United Kingdom
Str. Armeneasca 28/1, office 1, Chisinau MD-2012, Republic of Moldova, Europe
Printed at: see last page
ISBN: 978-620-8-01404-9

I dedicate this work to my family who, in the face of such simplicity, supported me and were the fundamental foundation on this journey, who constantly supported me in my quest for fulfilment and became my motivation, encouraging me, even if indirectly, every day up to this point. To all of you who have been, are and always will be my family, my thanks for all the opportunity, motivation and understanding to realise this dream.

ACKNOWLEDGEMENT

I would firstly like to thank God for providing many opportunities for this achievement to become a reality.

To my parents for every sacrifice, encouragement and understanding of my choices, for all the support I could feel even from afar, because it was you and through you that all my days made sense.

To Prof. Raymison Rodrigues Cardoso for the opportunity, dedication, patience and goodwill in each orientation for the elaboration of this work. And to Prof Dr Werner Farkatt Tabosa for his excellent performance in coordinating the TCC of the civil engineering course at UNIRN.

To my friends who have supported me, encouraged me and been with me on this journey, because they are the best I could ever have.

"We thank you, O God, we thank you, for your name

is near; everyone speaks of your marvellous deeds.

Psalm 75:1

SUMMARY

The aim of this monograph is to develop a study on the economic and environmental feasibility of using waste tyres in the asphalt paving of a highway and an avenue in Natal/RN. In order to carry out this study, it was necessary to gain an insight into the current situation regarding the collection of scrap tyres, rubber asphalt and conventional asphalt and last - and not least - economic competitiveness, environmental development and road development. A case study was developed comprising an environmental analysis of the use of waste tyres in asphalt paving works, an economic evaluation of a project using rubber asphalt and the economic viability of the transition from conventional asphalt to rubber asphalt. The development of the environmental analysis was based on articles on the environmental impacts caused by rubber asphalt technology. For inspection of BR-304/RN, the stretch from DIV CE/RN - 'ENTR BR-101(B) (COMPLEXO VIÁRIO DO 4° CENTENÁRIO - NATAL) TRECHO URBANO and Av. Prudente de Morais with the stretch R. Mossoró - Av. Alexandrino de Alencar for the types of pavement to be analysed using data from the Construction Reference Cost System (SICRO) and the Restoration and Maintenance Contract Project (CREMA) of the National Department of Transport Infrastructure (DNIT), as well as data from the Municipal Department of Public Works and Infrastructure (SEMOV), from which we will obtain the necessary costs for the transition from Hot Mix Bituminous Concrete (CBUQ) to rubber asphalt. The conclusion is that, although the conventional asphalt method is the most widely used in Brazil, the rubber asphalt method is better at minimising costs, saving on maintenance and reducing environmental impacts.

Keyword: Rubberised asphalt. Waste tyres. Environment. Savings.

SUMMARY

CHAPTER 1

INTRODUCTION

The pavement is a multi-layered structure built on top of the earthworks and designed, technically and economically, to withstand the stresses of traffic and, more directly, climatic action. It can be classified into three types: flexible, semi-rigid and rigid, and the most widely used in Brazil is flexible. It is therefore known that the surfacing layer must be impermeable and resistant to contact stresses, which vary according to the load.

One of the aims of new research worldwide and in all areas is to combine technology with environmental conservation, bringing innovation together with the reuse of waste and residues thrown into nature. With the growth of the vehicle fleet in Brazil, there has been a huge increase in tyre production, which has consequently led to a greater number of unserviceable tyres being disposed of incorrectly in the environment, thus causing a major environmental problem, as they were disposed of inappropriately in rivers, streets, rubbish dumps, forests and became a major problem for the environment and, of course, for society.

Currently, these unserviceable tyres are collected from the streets by an urban services company and sent to an appropriate place, where they will be reused correctly and will not cause harm to the population. In Natal, 500 unserviceable tyres are collected every day. The Companhia de Serviços Urbanos de Natal (Urbana) has an agreement with Reciclanip, which is an organisation dedicated exclusively to the collection and disposal of tyres in Brazil. Once collected, the tyres are taken to a cement factory to be reused.

Another solution for these tyres emerged in the United States and has been applied in some other countries, including Brazil. It is the use of dust from tyre rubber to complement the composition of asphalt, in which around 20% of the dust is added to the asphalt, thus making asphalt-rubber. There are already stretches of road with this type of paving in Brazil, such as the mountainous stretch of the Anchieta and Bandeirantes motorways, both located in the state of São Paulo. On average, 600 tyres are needed for every kilometre of paving. This practice makes the production of asphalt more expensive, an average of 15% more in production costs, but in the long term the benefits are numerous. Rubber asphalt is superior in quality to conventional asphalt, it has greater resistance which increases its durability and reduces the number of maintenance jobs, its high viscosity reduces the possibility of cracks, it avoids the risk of aquaplaning on rainy days, and it adheres better to the asphalt aggregate. It should also be remembered that this practice will also reduce the number of tyres improperly disposed of in Natal, thus reducing damage to the environment.

In the face of market storms, one factor that remains in evidence is the importance of asphalt paving, which gives us the right to come and go safely, more comfort, access to health and education, and we mustn't forget that today, roads are much more important for the country's economy because they are the arteries in Brazil most used for transporting goods. So why not invest in new road technologies? Bearing in mind that this greater investment has a long-term return for countless segments of the economy.

1.1 OBJECTIVES

1.1.1 General

To present a study on the economic and environmental feasibility of using scrap tyres for asphalt paving on a highway and an avenue in the city of Natal/RN.

1.1.2 Specific objectives

Carry out an environmental analysis of the use of waste tyres in asphalt paving works;
To study the economic viability of the transition from conventional asphalt to rubberised asphalt;
Carry out an economic assessment of a project using rubberised asphalt on Prudente de Morais Avenue in Natal/RN;
>- Quantifying the tyres needed to use rubber asphalt on the Prudente de Morais Avenue construction site in Natal/RN.

1.2 BACKGROUND

The north-eastern region faces a need for technological innovation, as does the whole of Brazil. There is a high consumption of tyres in Brazil due to the need for vehicles to move around more quickly, comfortably and safely, as road transport is still the most widely used mode. In Natal, according to data corroborated by Reciclanip, currently 500 unserviceable tyres are collected every day in the city, which generates a backlog of tyres that need to be disposed of in an environmentally correct way.

In Natal, the number of tyres collected is being made available to another state. It is known that if this collected waste were to be used for the municipality itself, it would be applied in various ways that have already been found, such as: car mats, industrial flooring, decoration, etc. And among these, one that is not new in Brazil, but would be new in Natal because it does not yet have an applicability, which is rubberised asphalt. This technology would bring a new lifestyle, improvements to health, possible improvements to the impact on the environment and possible improvements to the economy.

Rubber asphalt technology is made up of asphalt binder and 20% ground rubber with a granulometry of 0.6 to 1.2 mm. It is worth pointing out that using a percentage of rubber higher than this can cause the product to lose quality, reducing its resilience, and using a granulometry lower or higher than that mentioned above may not achieve the desired result in terms of applicability and may not guarantee good stability of the mixture. The asphalt binder and ground rubber should be mixed at a temperature above 177° for 20 minutes. The temperature used by the concessionaires is 190°. Given this information, the municipality of Natal/RN can acquire this technology because there is no difference in applicability between regions and it has been successfully applied in São Paulo, Minas Gerais and Paraná.

The importance of this work is justified by the belief that the use of scrap tyre rubber can help preserve the environment and that, in the long term, it can contribute economically and bring safety to the population. It can also minimise the traumatic effects of solid waste on the environment.

CHAPTER 2

METHODOLOGY

This work aims to study the economic and environmental viability of applying waste tyres to asphalt paving in the municipality of Natal. For the development of this work, bibliographical research was carried out on the subject in question, with the aim of understanding the different variables for the application of this technology in Natal, knowing that it does not have any industry in this line of action, and also looking at the potential applicability of this economically and environmentally sustainable method according to the conditions of this city, which may or may not receive it.

The economic analysis of certain factors, such as the cost of producing asphalt mix, whether conventional or rubber, as well as the durability forecast for both, will provide the theoretical basis for the case study, which will cover the BR 304/RN motorway, section DIV CE/RN - 'ENTR BR-101(B) (COMPLEXO VIÁRIO DO 4° CENTENÁRIO - NATAL) *TRECHO URBANO* and Sub-section: ENTR RN-042/263 (ANGICOS) - 'ENTR BR-226(A) from Segment 149.50 km to km 281.00 with Extension of 131.50 km in Lot 2 used as reference with the following geometric characteristics, being 37.6 km long with 7 m wide to 3 cm thick, 5.3 km long with 7 m wide to 4 cm thick, 14.8 km long with 7 m wide to 5 cm thick and both with a density of 2.425 t/m^3 . Also for the development of the economic analysis, this work will include Av. Prudente de Morais with the stretch R. Mossoró - Av. Alexandrino de Alencar using 347.40 m long with a width of 0.12 m for a thickness of 5 cm and a density of 2.425 t/m^3 . The environmental analysis will provide the theoretical basis for part of the case study which, in turn, will contain comparative environmental parameters from conventional asphalt to rubberised asphalt, such as: the environmental effects of this application on the water table, the impact on the population's health, etc. In addition to comparing the cost of implementing these different types of surfacing over time.

In order to carry out the case study for this work, we will also use data from the National Department of Transport Infrastructure's (DNIT) Work Reference Cost System (SICRO) and data from the Municipal Department of Public Works and Infrastructure (SEMOV) which, in its entirety, verified the costs necessary for the transition from Hot-Mix Bituminous Concrete (CBUQ) to rubberised asphalt using the chosen geometric characteristics as a reference, and will then be able to verify the economic viability. We will carry out an analysis using environmental data, which will provide results on the environmental viability of this transition and, finally, we will quantify the tyres needed to use the technology in question on the Prudente de Morais avenue in Natal/RN.

This methodology is based on bibliographical references, since any and all publications on the subject could contain indispensable information for a better understanding, as well as research into companies that have already used this alternative.

CHAPTER 3

LITERATURE REVIEW

3.1 TYRE COLLECTION IN BRAZIL

Among the solid waste produced by the population, tyres occupy a prominent role in the discussion of health and environmental impacts. Nowadays, landfill sites don't take tyres in one piece, as they are long-lived and can withstand constant impact, making them difficult to dispose of. When they are compacted whole, tyres tend to return to their original shape and come back to the surface, causing movement in the soil (GOMES, 1993).

The tyre industry began its activities in Brazil in the 1920s. In 1960, the National Tyre Industry Association (ANIP) was founded, a non-profit organisation whose aim is to defend the interests of the sector. In 1999, the National Programme for the Collection and Disposal of Unserviceable Tyres was launched, based on Resolution 258/99. In 2007, Reciclanip was created to strengthen the initiatives already in place, following the management model of European companies with extensive experience in the collection and disposal of unserviceable tyres. Currently, the resolution followed is CONAMA's 416/09 (IBAMA, 2017).

Reciclanip deals exclusively with the collection and recycling of unserviceable tyres. It represents tyre manufacturing companies in Brazil. Its activities are supported by investment from companies. In this respect, Reciclanip differs from European companies, since in other countries companies are paid by the various players in the production chain to cover operating costs and guarantee the disposal of unserviceable tyres. It is also responsible for setting up collection points with partners, managing the product's reverse logistics and promoting new destinations. The targets are calculated and monitored in tonnes. To calculate the tonnage target, a wear factor of 30% is applied to the weight of the new tyre (IBAMA, 2017).

Firstly, according to the National Environment Council, post-consumption responsibility was placed on the manufacturer and importer of the product and, later, co-responsibility for collection was placed on distributors, retailers and consumers. At present, Brazil has collection points in all states and the Federal District, initially set up in partnership with town halls in municipalities with more than 100,000 inhabitants.

The figure of more than 4.1 million tonnes of unserviceable tyres collected is equivalent to 821 million passenger car tyres. Since the programme began in 1999, tyre manufacturers have invested R$898.8 million in the initiative. "Unserviceable tyres are those that can no longer be used for circulation or retreading," explains Reciclanip manager Cesar Faceio.

Since 2010, the year of the IBAMA report, Reciclanip has been exceeding its logistics target. In 2015, for example, it exceeded the reverse logistics target set by IBAMA, achieving 101.7% collection and correct disposal of tyres. Brazil has an environmental liability of 179,500 tonnes of unserviceable tyres for which importers have not met their target. For this reason, one of ANIP's tasks is to seek measures from the federal government to resolve these liabilities, suggesting that import licences only be granted to those who have an environmental management contract or who pay a fee for the environmental disposal of unserviceable tyres in a volume corresponding to the import (IBAMA, 2017).

According to a survey carried out by the National Confederation of Industry (CNI) in partnership with consultancy firm LCA, the tax levied on the collection, sorting, transport and recycling of tyres cost the sector R$9.3 million in 2013 and R$12.1 million last year.

ANIP explains that the exemption of solid waste, in addition to encouraging its use as a raw material, helps to lower the cost generated by the onerous reverse logistics of tyres, in the same way as other products. For Faceio, the total exemption of taxes and duties - including ICMS for the transport of whole or shredded unserviceable tyres - will make the raw material more competitive with the virgin product in the manufacture of artefacts, as well as increasing its use on Brazilian streets and roads through rubber asphalt. "Our intention is to adopt a system similar to Europe's, where the cost is covered by a fee paid by the consumer when they buy the tyre, which shares responsibility across the entire chain and also independent importers, some of whom don't collect it," says ANIP's general manager.

Compliance with the national disposal target remained above 90 per cent. Manufacturers and importers of new tyres achieved 96.66% of the target. Graph 1 shows the historical evolution of the national disposal target.

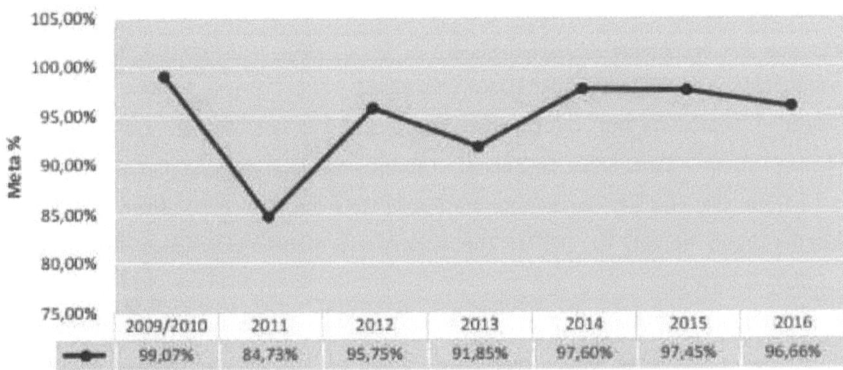

Graph 1 - Percentage of Compliance with the National Destination Target

Source: Ferreira (2017).

According to Brazilian Standard 10.004 of the Brazilian Association of Technical Standards (ABNT), tyres are considered class III waste (inert waste) in relation to the risk of their degradation in the environment. There are many environmental problems caused by the improper disposal of tyres:

> Its reuse in landfills is a dubious ecological practice, both because of the difference in degradation time compared to other materials used, and because it causes a hollow in the mass of waste, causing landfill instability.

> Due to its chemical composition, tyres are easy to burn and can cause fires that are difficult to control in landfills, releasing toxic and carcinogenic gases and oils into the soil, air and groundwater. For this reason, the open burning of tyres is prohibited in Brazil and in most countries around the world.

> In addition to the risks already mentioned, tyre carcasses exposed to the open air have the sanitary inconvenience of serving as a breeding ground for insects and rodents, making it difficult to control diseases such as dengue, malaria and yellow fever, thus also posing a serious threat to public health. There are suspicions, for example, that the *Aedes albopictus* mosquito, one of the vectors of the dengue and yellow fever viruses, has entered national territory in shipments of used tyres from the USA and Japan.

3.1.1 Tyre collection in Natal

The collection of tyres in Natal took hold after the Ministry of Health launched a national mobilisation from 19 to 23 September 2015, which continued in Rio Grande do Norte with the support of the State Secretariat for Public Health, through the Situation Room and Regional Health Directorates, as well as the Municipal Secretariats for Health and Urban Cleaning Services.

Reciclanip took part in the action. The aim of the Ministry of Health was to eliminate possible breeding grounds for the Aedes aegypti mosquito in tyres, which can accumulate water and encourage the proliferation of the mosquito (DJAILDO, 2016).

Collecting, separating and correctly disposing of 1,100 tonnes of waste every day is the challenge facing Urbana - Companhia de Serviços Urbanos de Natal. A large part of this task is carried out by Natal City Hall with funds from the Public Cleaning Fee (TLP), which is collected together with the Urban Property Tax (IPTU). The monthly cost of these activities is close to R$13 million (MESQUITA, 2016).

Natal produces 800 tonnes of municipal solid waste or household waste every day. Approximately 45% is dry waste (paper, plastic, cardboard and metals) and 55% is wet waste (organic matter). "We send this type of waste, after selectivity (which isn't all of it yet, but a large part) to Braseco, which is the landfill to dispose of it correctly," says Thiago Mesquita - Urbana's operations director. As well as household waste, the urban cleaning company has to deal with other types of material. "We also collect Special Solid Waste. There are three types: tyres, prunings and rubble. The tyres are stored in a shed we have in Alecrim (Figure 1), which are then sent for recycling. The prunings and rubble we collect are used to recover degraded areas from mining activities, i.e. those ditches or giant holes. We are currently using this material to reclaim a degraded area in the Guarapes neighbourhood," says the director. The inert material, which doesn't contaminate the soil, is deposited at these reclamation sites in order to restore the topography of the areas.

Urbana follows a daily collection route that covers the four regions of Natal. After collection, the tyres are taken to a shed in the Alecrim district and then transported to the capital of Paraíba. As well as saving money, since the tyres are used as fuel in a cement factory located in the city of João Pessoa, the measure contributes to the preservation of the environment and prevents the proliferation of the dengue mosquito. "The National Association of Tyre Industries, through Reciclanip, collects the tyres every fortnight for cement production in the capital of Paraíba. The old tyres are used to feed the kilns," explained Bosco Afonso, Urbana's CEO. The companies registered by Urbana keep the tyres for collection, and the cost of this collection is borne by the Companhia de Serviços Urbanos de Natal. In 2016 more than 142,000 tyres were collected at the shed, and in May 2017 there were already 47,000 tyres. For Urbana's president, Cláudio Porpino, the company's initiative is important. Currently 500 unserviceable tyres are collected every day in Natal (MESQUITA, 2016).

Figure 1 - Shed for storing waste tyres in Natal/RN

Source: Urbana (2011).

3.1.2 Collection points

The programme is carried out through partners, in most cases city halls, which provide land in accordance with specific safety and hygiene standards. This site is used to collect and store material from a variety of sources, such as tyre repair shops, retailers and citizens themselves. The person in charge

the collection point informs Reciclanip of the need to collect the material when it reaches 2,000 passenger tyres or 300 truck tyres. From then on, the company schedules the removal of the material with the agreed transporters. It is important that the collection point area is covered and protected to prevent water from accumulating or unauthorised people from entering. Figure 3 shows the number of collection points registered in each state on a map. In 2016, 1,723 collection points were registered, 932 of which were located in municipalities with a resident population of more than 100,000 inhabitants, leaving 13 municipalities with this characteristic without any declared collection points. Figure 2 shows the full list of municipalities and collection points. Figure 3 shows the complete list for Rio Grande do Norte (IBAMA, 2017).

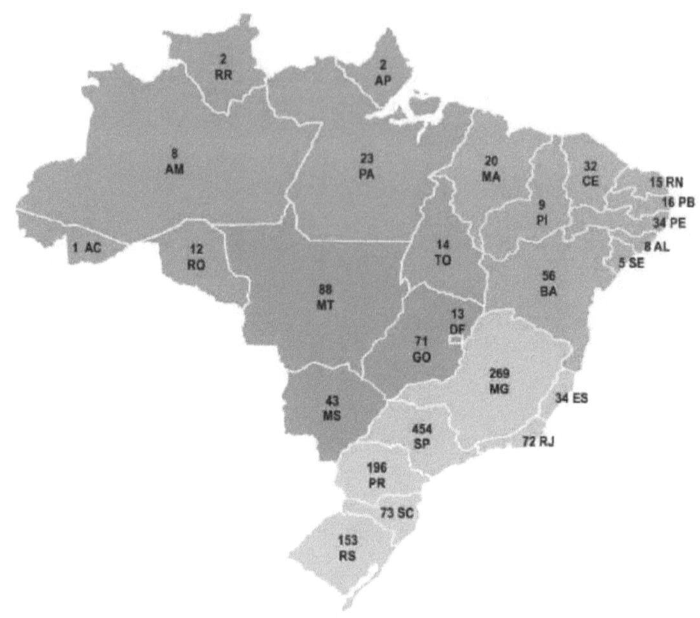

Figura 2 - Waste Tyre Collection Points by State (2016)

Source: Ferreira (2017).

RIO GRANDE DO NORTE

Município	Estado	ENDEREÇO	Capacidade (unidade)
CAPACIDADE TOTAL DO ESTADO			24.400
CAICO	RN	RUA VALDIR EPAMINONDAS LOPES, 635 – WALFREDO GURGEL	2000
CARNAUBA DOS DANTAS	RN	RUA JOSÉ MATIAS, 22 – CENTRO	2000
CURRAIS NOVOS	RN	RODOVIA BR 427, KM 02 – PARQUE INDUSTRIAL	2000
MOSSORO(*)	RN	AVENIDA CUNHA DA MOTA, 28 - CENTRO	100
MOSSORO(*)	RN	RUA JOSÉ DAMIÃO, 225 – SANTO ANTÔNIO	2000
NATAL(*)	RN	AVENIDA GOVERNADOR ANTONIO DE MELO E SOUZA, 2038-A	4000
NATAL(*)	RN	AVENIDA OESTE S/N – CIDADE NOVA	2000
NATAL(*)	RN	PRAÇA JOSÉ DA PENHA, 145 – RIBEIRA	100
NATAL(*)	RN	RUA MANDEL MASCARENHAS HOMEM, 105 – SAN VALLE	2000
PARNAMIRIM(*)	RN	AVENIDA PILOTO FERREIRA TIM, 803	100
PARNAMIRIM(*)	RN	AVENIDA PILOTO PEREIRA TIM, 848 – PARQUE DE EXPOSIÇÕES	2000
PARNAMIRIM(*)	RN	RUA FELIZARDO MOURA, S/N – LIBERDADE	2000
PARNAMIRIM(*)	RN	RUA GASPAR HENRIQUE CRUZ, 265	100
SANTA CRUZ	RN	RUA OLAVO BILAC, S/N	2000
SAO GONCALO DO AMARANTE	RN	LOTEAMENTO OLHO D'ÁGUADO CARRILHO	2000

Figura 3 - Registered Unserviceable Tyre Collection Points/RN - 2016
Source: Ferreira (2017).

3.2 RUBBERISED ASPHALT IN THE WORLD

Rubber asphalt is an asphalt binder that adds innovation, economy and sustainability to the pavement, conceived through a large investment in research for a product with high durability and safety, which brings with it respect for the environment and future generations (MORILHA JÚNIOR, 2016).

In the world, asphalt-rubber began in the 1940s, when the *U.S. Ruber Reclaiming Company* introduced to the market a product composed of asphalt material and recycled devulcanised rubber called *RamflexTM* (WICKBOLDT, 2005).

In the 1960s, Charles H. MacDonald, considered the father of rubberised asphalt in the United States, developed a highly elastic material to be used in asphalt pavement maintenance in 1963. The product consisted of asphalt binder and 20% ground tyre rubber (0.6 to 1.2 mm), mixed at 190°C for 20 minutes, to be used in patches known as

bandaid (WICKBOLDT, 2005).

Several countries have adopted rubberised asphalt not just as a good ecological solution, but as a way of obtaining more resistant pavements at a reasonable price and with low maintenance requirements. In the United States, where it was invented, rubberised asphalt has been used for around 40 years and already has more than 70% of Arizona's road network covered in rubber dust. France began using asphalt-rubber binders in 1982 and, in six years, more than 3,000,000 m² of this material have been applied to different types of surfacing (MAZZONETTO, 2011).The material is characterised by a mixture of asphalt alloy modified by crushed tyre rubber and hot compacted, a procedure used to increase the bond between the particles for commercial-scale production. Figure 4 shows a schematic and simplified model of rubber asphalt production.

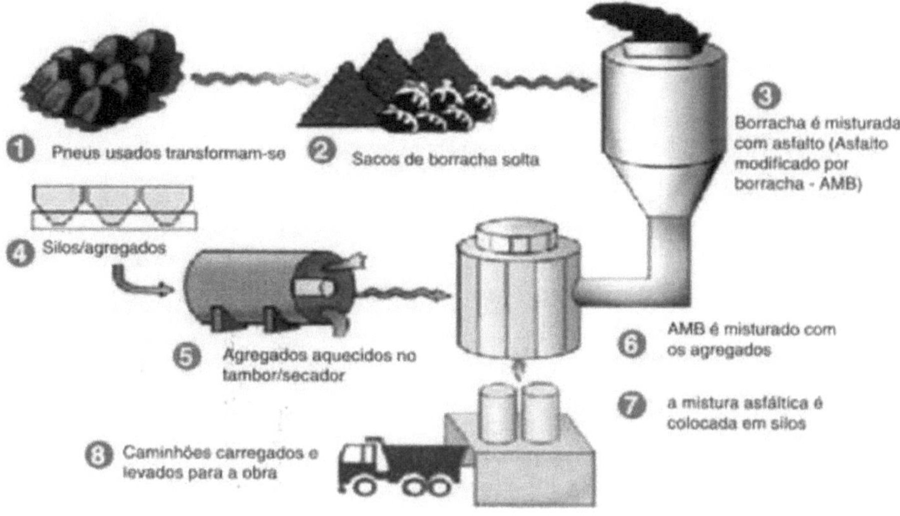

Source: Figueiredo (2013).

3.2.1 Rubberised asphalt in Brazil

The first steps towards the use of rubberised asphalt in Brazil were taken with the creation of Article 2 of Resolution 258/99 of the National Environment Council (CONAMA). The article prohibits the disposal of tyres in the environment, including open burning. This led to the need to redirect the disposal of unserviceable tyres. The year 1999 marked the start of the first AMB research in Brazil. A partnership between GRECA and the Paving Laboratory of the Federal University of Rio Grande do Sul (LAPAV) ensured progress in the results. The studies focus on a technology that uses rubber to improve the properties of ordinary asphalt. The first application in Rio Grande do Sul took place in August 2001. GRECA asphaltos is proud of both its pioneering spirit and the evolution ECOFLEX has undergone in its more than 10 years of existence. Today the product's technology is in its 3rd generation[a] , a differential that emphasises the reduction of CO2 and NO2 emissions from fuel consumption during operation. 2005 was the year of searching for special products. In all, more than 400 km of roads were paved with ECOFLEX during the year. The South and Southeast regions stood out in terms of demand for the product. The preference for Asphalt-Rubber is gaining ground after the product was standardised by bodies such as DER/PR and DEINFRA/SC. In December 2006, nine Brazilian technical papers stood out at the nationally important event: Asphalt Rubber 2006 - *The road to success.* During the year, works by the Ecovias dos Imigrantes Concessionaire on the Anchieta and Imigrantes motorways stood out (Figure 5). In the same year, the Paving Laboratory of the UFRGS School of Engineering (LAPAV) finalised the research entitled "Comparative Study of the Performance of a Resurfacing Using Asphalt-Rubber on a Flexible Pavement". The study stands out as one of the most important on Asphalt-Rubber.

Figura 5   - Application of Rubber Asphalt, Imigrantes Motorway, in SP Capital
Source: Mazzonetto (2011).

In 2007, DER/SP presented a technical standard for rubberised asphalt works. As a result, applications in the state of São Paulo increased significantly. At the same time, a technical partnership with South African engineers resulted in the first surface treatment project using Brazilian technology equipment. Brazil stands out for its numerous successful applications (Figure 6). The 3ª generation of ECOFLEX was launched in 2010, representing technological advances in the product that are reflected in economic and ecological factors. These improvements reduce the consumption of fuels used to machine the product. The environmental impact was reduced due to lower emissions of toxic gases. During the year, large-scale construction projects were carried out all over Brazil by organisations such as DNIT, DER in Paraná and São Paulo, DEINFRA in Santa Catarina and various concessionaires such as SPVIAS, Grupo EcoRodovias and AUTOBAN. GRECA[1] ended the year contributing more than 650 kilometres of roads paved with ECOFLEX. Ecological Asphalt reached a historic and representative milestone in 2012. With more than 5,000 kilometres paved and 5 million tyres removed from nature. These figures are due to the fact that for every kilometre paved with Asphalt-Rubber, around a thousand waste tyres are used. In 2017, the expectation is that after 9,000 kilometres of paved streets, roads and highways, the figure will reach 10,000,000 tyres consumed in the production of asphalt-rubber in Brazil.

---

[1] José Antonio Antosczezem Junior, 2012, historical article made available by Greca Asfaltos.

Figura 6 - RJ-122 Motorway Restored with Rubber Asphalt

Source: Mazzonetto (2011).

3.3 CONVENTIONAL ASPHALT

Flexible paving is one of the most widely used solutions in the construction and rehabilitation of urban roads, streets and motorways. According to data from the Brazilian Association of Asphalt Distributors (ABEDA), the paving system is made up of four main layers: asphalt base coat, base, sub-base and sub-base reinforcement. Depending on the intensity and type of traffic, the existing soil and the useful life of the project, the surfacing can consist of a bearing layer and intermediate or bonding layers. But in the most common cases, a single layer of asphalt mixture is used as the surfacing (NAKAMURA, 2011).

Asphalt can be manufactured in a specific plant, fixed or mobile, or prepared on site. In addition to the method of production, surfacings can also be classified according to the type of binder used: hot with the use of asphalt concrete, or cold with the use of Oil Asphalt Emulsion (EAP) (NAKAMURA, 2011).

CBUQ is the most widely used in Brazil. It is the product of mixing aggregates of various sizes and asphalt cement, both heated at previously chosen temperatures (Figure 7), depending on the viscosity-temperature characteristic of the binder (BALDO, 2007).

Figure 7 - CBUQ Caking as a Function of Viscosity-Temperature

Source: Asfalto (2017).

3.3.1 Conventional CBUQ asphalt

Hot-mix asphalt concrete (HMA) or also known as hot-mix asphalt concrete (HMA) can be considered the most common and traditional hot-mix asphalt mixture used in the country, used in the construction of pavement surfaces, including road surfaces and bonding layers (BALBO, 2007).

CBUQ is a flexible coating, but its choice depends on the nature of the work and the equipment available. It has properties and qualities such as: impermeability, adhesiveness, binder, durability, the possibility of working at different temperatures and an advantageous price (SENÇO, 2001, v. 1).

According to Balbo (2007), CBUQ is obtained by mixing and homogenising mineral aggregates, fine filler material and petroleum asphalt cement. It is therefore made from an elaborate hot mix and mixing plant. For its production, the aggregates must be dosed correctly, perfectly dry, through the heating drum, where their temperature is raised to a value compatible with the machining of the Petroleum Asphalt Cement (PAC) to avoid a drop in the temperature of the final mixture, which must reach the tarmac at between 140 and 145°C.

3.4 TECHNICAL COMPARISON BETWEEN CAP AND AB

With regard to the characteristics of conventional asphalt and rubberised asphalt, one of the relevant aspects is the formation of cracks and fissures in pavements. According to Bernucci et al, (2007) pavements with rubberised asphalt are more resistant to cracking and permanent deformation (wheel tracks). This is possible because the asphalt mix acquires some of the elastic capacity of the rubber and is thus able to deform as vehicles pass by and return to its initial position, thus reducing undesirable deformations. (BERNUCCI et al., 2007).

An important factor is that rubberised asphalt, due to some rubber substances such as *carbon* black (b/ack *carbon)*, protects the asphalt against chemical wear resulting from the pavement's exposure to infrared and ultraviolet rays, which are very intense in the country, thus preventing premature ageing of the asphalt (BERNUCCI et al., 2007).

Another factor in favour of using rubber asphalt is that it allows for the construction of rough, porous and self-draining pavements, which reduces the aquaplaning effect caused by the accumulation of water on the road (RODOVIAS; VIAS, 2012).

There are other factors in which rubberised asphalt is better than conventional asphalt. Rubber asphalt provides better grip, reduces noise by up to 85%, reduces ageing and thermal susceptibility, and increases resistance to fatigue and wheel track formation, as well as providing an appropriate final destination for scrap tyres, thus helping the environment. Although the rubberised asphalt binder is more expensive, this cost is offset by the low maintenance costs over the years (GRECA..., 2003).

According to Zanzotto and Svec (1996) and the Asphalt Rubber Pavement Association (apud GRECA..., 2003) the binder modified by granulated tyre rubber, or simply rubber asphalt, has the following characteristics:

> Reduced thermal susceptibility: mixtures with rubberised asphalt binder are more resistant to temperature variations, i.e. their performance at both high and low temperatures is better when compared to pavements built with conventional binder;
> Increased flexibility due to the higher concentration of elastomers in tyre rubber;
> Better adhesion to aggregates;
> Increased pavement service life;
> Greater resistance to ageing: the presence of antioxidants and carbon black in tyre rubber helps to reduce ageing due to oxidation;
> Greater resistance to crack propagation and wheel track formation;
> Reduces pavement thickness;
> Provides better tyre-pavement grip;
> Traffic noise reduction of between 65 and 85 per cent.

According to Bernucci et al. (2007), as a constructive disadvantage, high rubber contents lead to high viscosities, which compromise the workability of the asphalt mix. In addition, machining temperatures are slightly higher than for conventional asphalt.

The performance of rubber asphalt and conventional asphalt can be assessed using various laboratory tests such as: Softening Point; Ductility; Adhesiveness; Viscosity; Penetration, among others (BERNUCCI et al, 2007).

3.5 RUBBER ASPHALT PLANT TECHNOLOGY

According to Wickboldt (2005), there are two methods of obtaining rubberised asphalt: dry or wet. In the dry process, the rubber is fed directly into the asphalt plant's mixer. In this case, the rubber is added to the mix as an aggregate. The transfer of important properties from the rubber to the binder is impaired, although it is possible to add improvements to the asphalt mix, as long as it is possible to obtain a homogeneous mix. In the dry process, the rubber granules account for 0.5 to 3.0 per cent of the aggregate mass. In the wet process, the rubber is previously mixed into the binder, permanently modifying it. In this way, the characteristics of elasticity and resistance to ageing are transferred more effectively to the conventional asphalt binder. In the wet process, tyre dust accounts for 15% to 20% of the binder's mass or less than 1.5% of the mixture's mass. Some rubber fragments are mixed with the heated asphalt cement, producing a new type of binder called Modified Tyre Rubber Pavement (PMB). In the wet process, the asphalt-rubber binder is obtained by adding the crushed rubber to the asphalt binder at temperatures between 175 and 200°C. After the reaction period, the product is stored in a tank, which must contain a mechanical system of constant agitation in order to keep the mixture dispersed, so as to avoid the deposition of particles that have not partially reacted with the asphalt binder.

According to Petrobras' asphalt pavement analyser, a device that measures fatigue and permanent deformation, the changes suffered by rubber asphalt after ten years of use are up to four times less, and the fatigue life is more than double, in some cases triple that of pavements made with conventional asphalt (CONCER, 2009).

The materials used in the mixtures to make rubber asphalt are: rubber from ground-up discarded tyres and petroleum asphalt cement, with the rubber used in the binder coming mainly from car tyres (ODA; FERNANDES JÚNIOR, 2001). Figure 8 shows a model of a rubber asphalt plant located in the state of São Paulo.

Figure 8 - Model of a Rubber Asphalt Plant
Source: Mazzonetto (2011).

3.6 BRAZIL'S ECONOMIC COMPETITIVENESS IN INFRASTRUCTURE

The Infrastructure factor was the only one to show a significant improvement in 2016 with a gain of 7 positions (46[a]), as shown in graphs 2 and 3. This factor has gone through several oscillations over time. The basic infrastructure sub-factor kept its indicators relatively stable and made gains relative to the other countries by climbing 5 positions. This gain was partly due to the improvement of 8 positions in the "access to water" variable, which in 2015, due to the water crisis, placed Brazil in last place in this indicator. Logistics management, distribution, energy infrastructure and rainwater transport continue to be among the worst in the world, with the country occupying the last positions in the ranking (logistics management: 60°; distribution: 60°C, energy infrastructure: 58°C, rainwater transport: 59[a]). In this block of basic infrastructure variables, apart from the improvement in access to water, the only variable to show significant improvements was the quality of air transport, which fell from 59[a] in 2015 to 53[a] in 2016. Far below the country's needs, but consistent with privatisation and investment in Brazil's airport infrastructure (PROF. CARLOS ARRUDA, 2017).

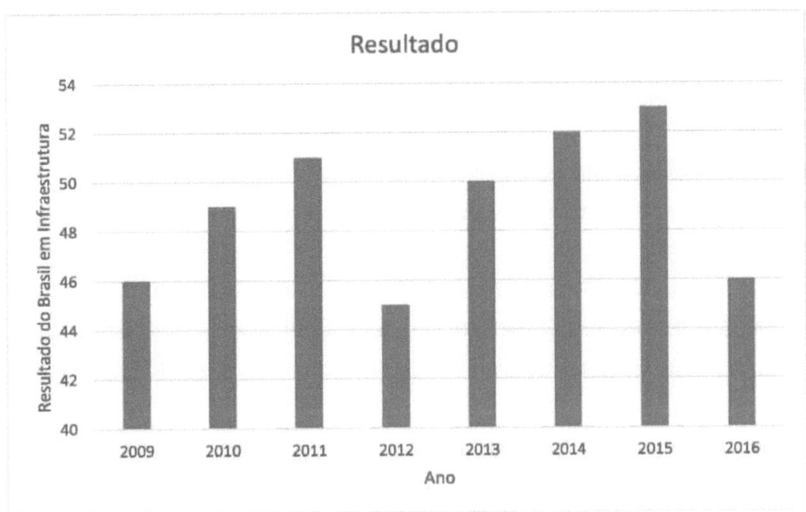

Graph 2 - Brazil's results in Infrastructure (2009-2016)
Source: IMD(2017).

With the restoration of the flow of federal investments earmarked for road transport and aimed at widening, modernising and conserving highways, various works in all the states were resumed from 2003 onwards.

This portfolio consists of widening and upgrading works covering 3,337 km, as well as paving and building highways totalling 5,328 km, benefiting all five regions of the country. Continuing the maintenance of federal highways, the coverage contracted for these services exceeds 48,000 km. Road maintenance services aim to ensure good road conditions, provide safety for users and reduce transport costs. Of this contracted length, 18,700 km are under contract[2] .

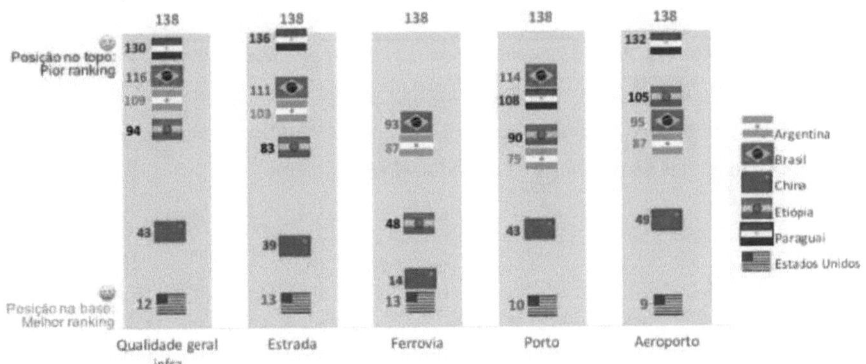

Graph 3 - Global Competitiveness Ranking: Brazil in Infrastructure

---

[2] Report made available by the Growth Acceleration Programme - PAC

3.7 ECONOMIC ROAD DEVELOPMENT IN BRAZIL

The general state of Brazilian roads worsened in 2017 and the classification of regular, bad or terrible is close to two thirds of the total analysed by the National Transport Confederation (CNT) in the "CNT Road Survey 2017".

"The decline in the quality of Brazilian roads is directly related to a history of low investment in road infrastructure and the economic crisis of recent years" (CNT, 2017), said CNT president Clésio Andrade. Data compiled by the organisation shows that, in 2016, public investment in highways was R$8.61 billion.

When broken down by region, the North has the worst results, with 81.1 per cent of results being fair, bad or terrible. Next comes the Midwest, with 65.4%, and the South (61.7%). Completing the list are the Northeast, which had 61.5% of its roads rated as regular, bad or terrible, and the Southeast (51.5%). Brazil has only 12.3% of its road network paved.

The need for Brazil's economy to recover in the coming years will increase the demands for greater efficiency in the transport infrastructure and, above all, will reinforce the perception of the country's need for higher quality roads. In this context, ensuring the recovery and expansion of our road network is essential if we are to achieve social and economic growth on a permanent basis. After going through the worst recession in

Brazil needs to consolidate the economic recovery process seen in the second half of 2017, as can be seen in Graph 4. Expanding investment in infrastructure is the fastest and surest way to achieve a new cycle of sustainable development, with job creation and income distribution for all Brazilians. Overcoming the barriers imposed by deficiencies in transport and logistics infrastructure presupposes the recovery and expansion of the country's road network, through which the majority of people and a large part of national production pass (CNT, 2017)[3] . It can be seen in graphs 4 and 5 that the current state of the roads is due to a lack of investment.

---

[3] Clésio Andrade, President of the CNT.

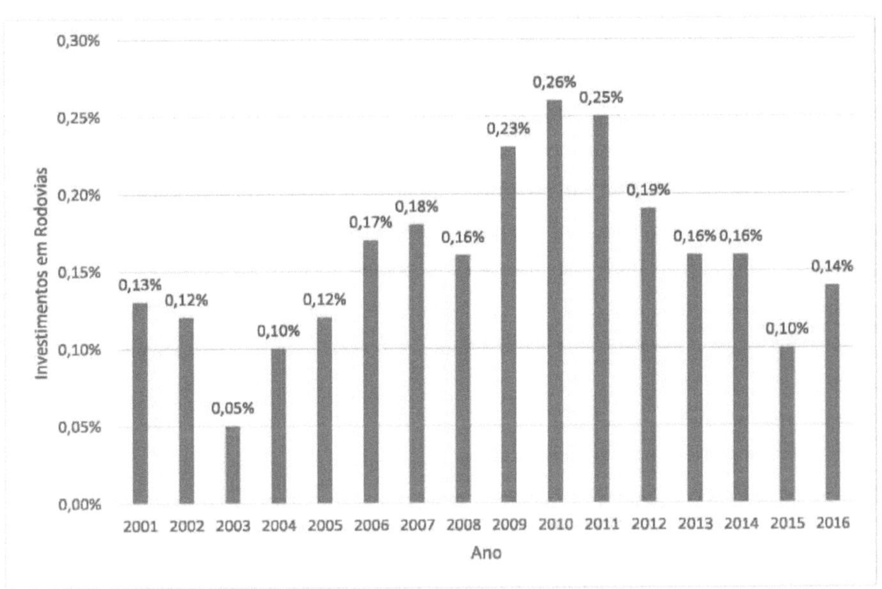

Graph 4 - Investment in Roads, Brazil - 2001 to 2016 (%)

Source: Adapted from CNT (2016).

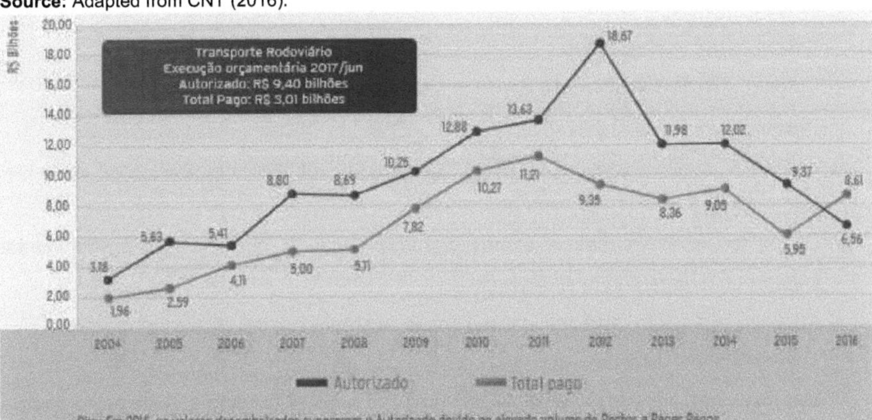

Graph 5 - Public Investment in Road Transport Infrastructure

Source: CNT (2016).

3.8 ENVIRONMENTAL DEVELOPMENT IN BRAZIL

Sustainable development is a concept designed to refer to the environment and the conservation of natural resources. Sustainable development is understood as the ability to utilise nature's resources and goods without compromising the availability of these elements for future generations (PENA, 2017).

This means adopting a pattern of consumption and utilisation of raw materials extracted

from nature so as not to affect the future of humanity, combining economic development with environmental responsibility. It is wrong to think that the latter are inexhaustible, as misuse could extinguish their availability in nature, with the exception of wind and sunlight, which are not directly affected by economic exploitation practices. The concept of sustainable development was officially declared at the United Nations Conference on the Human Environment, held in 1972 in Stockholm, Sweden, and therefore also called the Stockholm Conference. The importance of drawing up the concept at that time was to unite the notions of economic growth and development with the preservation of nature, issues that until then had been seen separately. Among the measures that can be adopted by both governments and civil society in general to build a more sustainable world are the following
world based on sustainability, we can mention:

> Reducing or eliminating deforestation;

> Reforestation of devastated natural areas;

> Preservation of environmental protection areas, such as reserves and riparian forest conservation units;

> Supervision by the government and the population of acts of environmental degradation;

> Adoption of the 3Rs policy (reduce, reuse and recycle) or the 5Rs (rethink, refuse, reduce, reuse and recycle);

> Curbing waste production and directing it correctly to reduce its impact;

> Decrease in the incidence of fires;

> Reduction in the emission of pollutants into the atmosphere, both from industrial chimneys and vehicle exhausts and others;

> Opting for clean sources of energy production that do not generate environmental impacts on a large and medium scale;

> Adoption of ways to raise political and social awareness of the above measures.

These measures are therefore viable and practical ways of building a sustainable society that does not jeopardise the natural environment either today or in the medium and long term (PENA, 2017).

What choice do we have? According to some economists, we need to gradually reduce the demand for raw materials by reusing and recycling materials and products. But this would only be a small step.

CHAPTER 4

CASE STUDY

The environmental analysis was drawn up after extensive research into some of the parameters that can alter the environment in a positive or negative way, directly or indirectly influencing the lives of the population of Rio Grande do Norte and more precisely the city of Natal/RN.

The economic analyses were carried out using data from the Sistema de Custos Referenciais de Obras (SICRO) of the Departamento Nacional de Infraestrutura de Transportes (DNIT) which, in its entirety, we obtained the necessary costs for the transition from hot-mix bituminous concrete (CBUQ) to rubber asphalt using as a reference the geometric characteristics found in the CREMA Programme Project for the BR 304/RN road section DIV CE/RN - 'ENTR BR-101(B) (COMPLEXO VIÁRIO DO 4° CENTENÁRIO - NATAL) *TRECHO URBANO* and Sub-stretch: ENTR RN-042/263 (ANGICOS) - 'ENTR BR-226(A) from Segment km to km 281.00 with an Extension of 131.50 km in Lot 2, then being able to verify the economic viability of the cost of producing asphalt mix also using data from the Municipal Department of Public Works and Infrastructure (SEMOV) for an evaluation on an avenue in the city of Natal/RN.

The Restoration and Maintenance Contract (CREMA) project was conceived and developed in 1999 as part of the Restoration and Decentralisation of Federal Highways Programme, and is the milestone for changing the management model for maintaining the federal highway network, as part of loan agreement 4188-BR signed between the Brazilian government and the World Bank. The basic scope of this project consists of a tender, location map, survey and evaluation standards, a catalogue of solutions and project preparation. CREMA-type programmes ensure that the good condition of the road is maintained for a period of two years (CREMA 1ª stage) or five years (CREMA 2ª stage) and with the solutions covered by the CREMA 2 programmeª stage, the useful life of the roads is up to 10 years.

4.1 ENVIRONMENTAL ANALYSIS OF THE USE OF SCRAP TYRES IN ASPHALT PAVING WORKS

Tyres play a role in our lives and this, in turn, is fundamental for our transport as passengers, for our economy and for that of the country when we understand that road transport is the most widely used mode for transporting goods. However, when they become unserviceable, they can also cause a series of health, environmental and visual problems. Some of the ways in which these problems can be corrected are through applications that are designed in such a way as to help and please the population, such as: furniture, shoes, roofs, decoration, vegetable gardens (these can be found in the Composting and Organic Vegetable Garden Pilot Project, coordinated by Prof. Dr. Werner Farkatt Tabosa). Werner Farkatt Tabosa, at the University Centre of Rio Grande do Norte) and, among many others, we also have the one that is being applied in the subject of this work, crushed rubber in asphalt mixtures which, as well as being ecologically correct, improves the performance of pavements, delays the appearance of cracks and reduces operating costs.

Every tyre will at some point become waste that is harmful to public health and the environment. In order to minimise these impacts, RECICLANIP, among other companies, has been collecting waste tyres and, in Natal/RN, around 100 tonnes of tyres are collected from the streets every year. A viable and economical way of reusing tyres is to use crushed rubber as a component of asphalt, as it improves the quality of the pavement and reduces costs, making it viable in the light of an economic environmental analysis.

With environmental legislation that makes companies responsible for the entire life cycle of their products in terms of their destination after delivery to their customers and the impact they have on the environment, what the population can still see is that reusing these scrap tyres reduces the risk of transmission of certain diseases through the aedes aegypti mosquito, such as dengue, chikungunya, zika, which is notable for its association with cases of microcephaly and yellow fever.

It is worth emphasising the existence of two concepts presented by Ormond (2004), which indicate the types of modifications to the environment attributed to man, which can be negative or positive. Changes that cause damage, destruction or degradation are called negative environmental impacts. However, it is also possible to develop actions to improve the situation, the scenario, regenerating degraded environmental areas or functions, thus indicating positive environmental impact through continuous environmental management.

For Specht (2007), from UFRGS, it's true that rubberised asphalt has to be heated to higher temperatures, but it's also a fact that it doesn't harm the health of workers or pollute more than conventional asphalt, as long as traditional precautions are taken. "As for the controversy that asphalt-rubber is more polluting, in the United States, which is the major user of this technology, this discussion has been outdated since the 1990s and the state of California is at the forefront in many environmental aspects (catalytic converters for cars, among others), using millions of tyres a year in asphalt-rubber," he analyses. Also according to Specht (2007), all asphalt plants, as a rule, have filters that prevent fuel smoke and stone dust from being released into the atmosphere and, in order to operate, they need an environmental licence (i.e. they are supervised by the public authorities).

4.2 ECONOMIC EVALUATION OF A RUBBERISED ASPHALT PROJECT

Overall, the use of asphalt binders modified by waste tyre rubber in asphalt mixtures for pavement resurfacing is a promising technique, as asphalt mixtures with rubber asphalt have lower maintenance values than those found in asphalt mixtures with conventional binders (GRECA, 2015).

We will therefore present analyses of a restoration project on the BR 304/RN with SICRO data, found in Table 1, in which the project specifies a layer of asphalt concrete with CAP-50/70 binder with different thicknesses for different homogeneous segments that have undergone road maintenance. We also present the budget for a coating with rubberised asphalt with a 30% reduction (this is possible and justified on the basis of international and national studies that indicate thickness reductions of up to 50%) in the thickness of the reinforcement and a density of 2.425t/m^3 (data taken from the CREMA base project) will be considered for each one.

Table 1 - CREMA Project data

DATA	4 cm	5cm
Extension	5.3 kilometres	14.8 kilometres
Width	7m	7m
Density	2.425 t/m^3	2.425 t/m^3

Source: Adapted from the CREMA Project (2017).

In the city of Natal thoro arc several plants to produce asphalt, so we take as a reference a plant with a production capacity of 80 tonnes per hour and, considering that it works 8 hours per day, we get a value of 640 tonnes per day, which taking into account 26 working days gives us a value of 16,640 tonnes per month. Using the data in table 1 for the homogeneous segment that is 4cm thick, we have a length of 5.3km and a width of 7m, which gives us the quantity of asphalt mix shown in table 2 below:

Table 2 - Quantities of Asphalt Mastic (4cm)

Surfacing - CBUQ (Conventional)	Surfacing - Rubberised Asphalt
5,300m x 7.0m x 0.04m x 2.425t/m^3 = 3,598.7t CBUQ asphalt mix	5,300m x 7.0m x 0.028m x 2.425t/m^3 = 2,519.09t rubber asphalt mix
3,598.7 tonnes	**2,519.09 tonnes**

Source: Adapted from the CREMA Project (2017).

For the 4cm thickness, after dividing the amount of asphalt mix by the monthly production of a plant, we have 0.2 months to apply CBUQ with conventional asphalt and 0.1 months to apply rubber asphalt - with a 30% reduction - and this results in 0.1 months of savings in fixed costs such as industrial facilities and the labour required to carry out this road surfacing solution.

Using the data quoted in table 1 for the homogeneous segment that is 5cm thick, we get a length of 14.8km and a width of 7m in which we find the following quantity of asphalt mix shown in table 3 below:

Table 3 - Quantity of Asphalt Mastic (5cm)

Surfacing - CBUQ (Conventional)	Surfacing - Rubberised Asphalt
14,800m x 7.0m x 0.05m x 2.425t/m³ = 12,561.5 tonnes of CBUQ asphalt mix	14,800m x 7.0m x 0.035m x 2.425t/m³ = 8,793.05t of Rubber Asphalt asphalt mix
12,561.5 tonnes	**8,793.05 tonnes**

Source: Adapted from the CREMA Project (2017).

In addition, for a thickness of 5cm we have 0.7 months to apply CBUQ with conventional asphalt and 0.5 months to apply rubber asphalt - a 30% reduction - and this results in 0.2 months of savings in fixed costs such as industrial installations and the labour required to carry out this application.

According to the cost principles adopted in table 4 below:

Table 4 - Principles Adopted

Price of PAC 50/70 (R$/tonne)	Price of Rubber Asphalt (R$/tonne)	Binder content (%)
1743,44	1790,25	5 CAP and 5.5 AB

Source: Adapted from SICRO (2017).

The price per tonne that pays for all the inputs and the application of the asphalt mix is as follows (Table 5):

Table 5 - Price per tonne

CBUQ - CAP50/70 (R$/tonne)	85,8997
CBUQ - Rubber Asphalt (R$/tonne)	121,1400

Source: Adapted from SICRO (2017).

According to table 5 above, the price of rubberised asphalt is 30% higher than the price of CBUQ - CAP50/70. This increase is equivalent to the costs of raising the machining temperatures of the asphalt mix and increasing the quality of compaction.Recapitulating the data found above, let's look at tables 6 and 7 below, which describe the costs of executing the coatings for each type of asphalt.

Table 6 - Description of Costs (4cm)

	DATA	CALC.	UND.	THICKNESS	CAP50/70	AB
1	Qty of CBUQ Asphalt Mastic (Produced)	-	tonne	4cm	3.598,70	2.519,09

	Machining/application cost per tonne of CBUQ applied	-	R$/tonne	4cm	85,8997	121,14
2						
3	Quantity of mass x Cost of Machining/Application	1X2	R$	4cm	309127,25	305162,56
4	Asphalt content	-	%	4cm	5	5,5
5	Asphalt cost per tonne	-	R$/tonne	4cm	1743,44	1.790,25
6	Cost Asphalt in CBUQ	1x4x5	R$	4cm	313.705,88	248.039,05
7	Total Cost of the Work	3+6	R$	4cm	622.833,13	553.201,61

Source: Adapted from SICRO and the CREMA Project (2017).

In this case, for 4cm, there is also a reduction in costs when using rubberised asphalt.

See the percentage below:

$$Rc = \frac{(622.833,13 - 553.201,61) \times 100}{622.833,13} = 11,2\%$$

Table 7 - Description of Costs (5cm)

	DATA	CALC.	UND.	THICKNESS	C AP 50/70	AB
1	Qty of CBUQ Asphalt Mastic (Produced)	-	tonne	5cm	12.561,50	8.793,05
2	Machining/application cost per tonne of CBUQ applied	-	R$/tonne	5cm	85,8997	121,14
3	Mass quantity x Machining/Application cost	1X2	R$	5cm	1079029,08	1065190,08
4	Asphalt content	-	%	5cm	5	5,5
5	Asphalt cost per tonne	-	R$/tonne	5cm	1743,44	1.790,25
6	Cost Asphalt in CBUQ	1x4x5	R$	5cm	1.095.011,08	865.796,68
7	Total Cost of the Work	3+6	R$	5cm	2.174.040,16	1.930.986,75

Source: Adapted from SICRO and the CREMA Project (2017).

It can therefore be seen that THERE is a reduction in costs when rubberised asphalt is used.

Below is an equation used to determine the percentage of this cost reduction (Rc).

$$Rc = \frac{(2.174.040,16 - 1.930.986,75) \times 100}{2.174.040,16} = 11,2\%$$

As we can see, the percentage cost reduction for the homogeneous segment using 4cm and 5cm thicknesses is 11.2%, both thicknesses taken from the CREMA Programme Project. In other words, 11.2% of the specified value can be saved by applying rubberised asphalt instead of conventional CBUQ.

4.3 ECONOMIC VIABILITY OF THE TRANSITION FROM CONVENTIONAL ASPHALT TO RUBBERISED ASPHALT

In order to carry out this study, the BR-304/RN motorway was selected, the segment linking the CE/RN border to Natal/RN. When inspecting BR-304/RN, the stretch from DIV CE/RN - ENTR BR-101(B) (COMPLEXO VIÁRIO DO 4° CENTENÁRIO - NATAL) TRECHO URBANO was chosen for the types of pavement to be analysed. The stretch of homogeneous segment covered with CBUQ was selected to compare the two different paving methods, showing that the costs would not be the same. To this end, the amounts invested in base preparation and maintenance costs found in SICRO were evaluated (Graph 6).

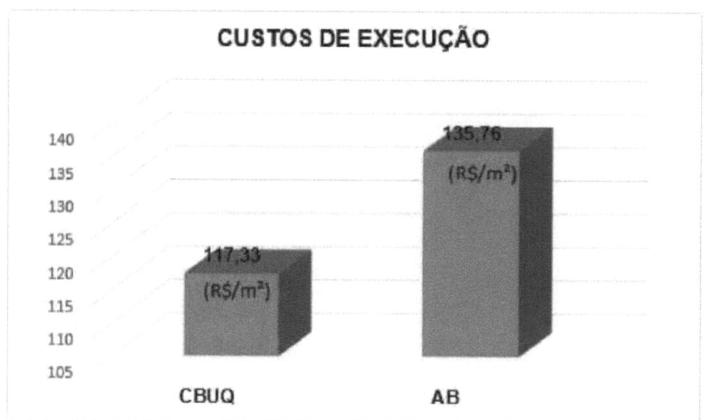

Graph 6 - Pavement Type Execution Data and Costs

Source: Adapted from SICRO (2017).

According to the data found in Graph 6, the cost of laying rubberised asphalt is 13.6% higher than the cost of laying conventional asphalt.

However, considering that after 5 years there will be different levels of wear on the roads and applying a percentage of the amount of maintenance on each stretch, 70% for CBUQ and 30% for AB, and also knowing that the cost of maintenance is R$147.69/m² according to SICRO data, then we will have new maintenance costs to apply (Graph 7).

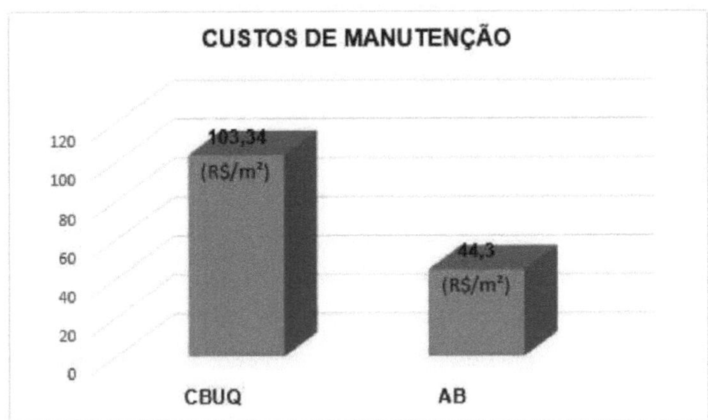

Graph 7 - New Maintenance Costs to be Applied

Source: Adapted from SICRO (2017).

Checking the maintenance values found in graph 7, it can be seen that the maintenance cost of conventional asphalt would be 57.13% more expensive than the maintenance cost of rubberised asphalt. In order to obtain a more accurate and complete comparison of the pavements, it is necessary to add up the maintenance and execution costs of each one, as shown in table 8 below:

Table 8 - Exact Comparison Values

PAVEMENTS	EXECUTION (R$/m $)^2$	MAINTENANCE (R$/m $)^2$	TOTAL (R$/m $)^2$
CBUQ	117,33	103,34	220,67
AB	135,76	44,3	180,06

Source: Adapted from SICRO (2017).

According to the data above, we have seen that the sum of the execution and maintenance values gives us a more realistic comparison. In addition, conventional asphalt is 18.4% more expensive than rubberised asphalt due to the fact that the CBUQ wears down more.

Summarising the data found above, let's take a look at Table 9 below, which describes the execution costs plus maintenance costs for each type of asphalt.

Table 9 - Summary of Costs for Comparing Types of Asphalt

	DATA	CALC.	UND.	CAP50/70	AB
1	Laying the floor	-	m	100	100
2	Pavement Execution Costs	-	R$/m²	117,33	135,76
3	Floor maintenance for 5 years		% m²	70	30

4	General cost of pavement maintenance	-	R$/m²	147,69	147,69
5	Percentage of pavement maintenance cost	3x4	R$/m³	103,38	44,31
6	Running costs plus maintenance	2+5	R$/m²	220,71	180,07

Source: Adapted from SICRO and the CREMA Project (2017).

As a result, it can be seen that there is a reduction in costs when rubberised asphalt is used.

4.4 ECONOMIC EVALUATION OF A PROJECT WITH RUBBERISED ASPHALT ON AV. PRUDENTE DE MORAIS IN NATAL/RN

Analyses of an avenue in the city of Natal will be presented, based on SEMOV data found in Table 10, in which the project specifies a layer of asphalt concrete with CAP-50/70 binder with a thickness of 5cm that has undergone road maintenance. We also present the budget for a coating with rubberised asphalt with a 30% reduction in reinforcement thickness and a density of 2.425t/m³.

Table 10 - SICRO data

DATA	5cm
Extension	347,40 m
Width	9,0 m
Density	2.425 t/m³

Source: Adapted from SEMOV (2017).

We continue with the reference of a plant with a production capacity of 80 tonnes per hour, which gives us a figure of 640 tonnes per day. This shows that

Using the data quoted in table 10 for the homogeneous segment that is 5cm thick, we have a length of 347.40m and a width of 9m, which gives us the quantity of asphalt mix shown in table 11 below:

Table 11 - Quantity of Asphalt Mastic (5cm)

Surfacing - CBUQ (Conventional)	Surfacing - Rubberised Asphalt
347.40m x 9.0m x 0.05m x 2.425t/m³ =	347.40m x 9.0m x 0.035m x 2.425t/m³ =
379.10 tonnes of CBUQ asphalt mix	265.40 tonnes of asphalt rubber compound
379.10 tonnes	**265.40 tonnes**

Source: Adapted from SEMOV (2017).

For the 5cm thickness used in the city of Natal/RN, we have 0.02 months to apply CBUQ with conventional asphalt and 0.01 months to apply rubber asphalt - with a 30% reduction - and this results in 0.01 months of savings in fixed costs such as industrial installations and the labour required to carry out this application.

Considering the principles adopted in table 4 and the prices per tonne in table 5, we have recapitulated the data found previously in table 12 below, describing the costs of carrying out the overlays for each type of asphalt with a thickness of 5cm.

Table 12 - Description of Costs (5cm)

	DATA	CALC.	UND.	THICKNESS	CAP50/70	AB
1	Qty of CBUQ Asphalt Mastic (Produced)	-	tonne	5cm	379,1	265,4
2	Machining/application cost per tonne of CBUQ applied	-	R$/tonne	5cm	85,8997	121,14
3	Mass quantity x Machining/Application cost	1X2	R$	5cm	32564,58	32150,56
4	Asphalt content	-	%	5cm	5	5,5
5	Asphalt cost per tonne	-	R$/tonne	5cm	1743,44	1.790,25
6	Cost Asphalt in CBUQ	1x4x5	R$	5cm	33.046,91	26.132,28
7	Total Cost of the Work	3+6	R$	5cm	65.611,48	58.282,84

Source: Adapted from SEMOV (2017).

It can therefore be seen that there is a reduction in costs when rubberised asphalt is used. Below is an equation used to determine the percentage of this Cost Reduction (Rc).

$$Rc = \frac{(65.611,48 - 58.282,84) \times 100}{65.611,48} = 11,17\%$$

It can be seen that the percentage cost reduction for the homogeneous segment using 5cm thickness is 11.17%, i.e. 11.17% of the specified value can be saved by applying rubberised asphalt instead of conventional CBUQ.

4.5 QUANTIFICATION OF TYRES REQUIRED FOR THE USE OF RUBBER ASPHALT ON THE AV. PRUDENTE DE MORAIS IN NATAL/RN.

In Brazil, some of the waste tyres are reused in various ways, after being ground down and separated from the other components of the tyre, especially the steel. Among the products that reuse rubber is mixing it with asphalt for use in paving, generating rubberised asphalt, which has important advantages. The waste tyres collected in Natal/RN number more than 500 a day and are sent to

João Pessoa/PB where they are burnt as fuel in the cement industry.

In order to apply rubber asphalt technology in Natal/RN, it was necessary to quantify the waste tyres collected daily, monthly and annually in the municipality itself.

In order to quantify the number of tyres needed to use rubber asphalt on the Prudente de Morais avenue construction site in Natal/RN, it was necessary to search for the number of waste tyres used for a given distance. Table 13 below shows the results for 3 consortia for a distance of 1 kilometre.

Table 13 - Unserviceable Tyres Required for a 1km Extension

DATA	Unserviceable Used
Consortium A	750 tyres
Consortium B	600 tyres
Consortium C	1000 tyres

Source: Lacerda (2016) and Mazzonetto (2011).

After performing a weighted average of the results obtained in Table 13, it was possible to obtain an average number of waste tyres needed for 1km, resulting in a figure of 783.33 waste tyres needed.

In order to obtain the amount of asphalt mix that can be produced with the material collected in Natal/RN, we had to make a calculation using the data in Table 14 below.

Table 14 - Data for calculating KM produced

Extension	Unserviceable Necessary
1km	783.33 tyres
X	15,000 tyres

Source: Adapted from SOUZA (2016).

Therefore, when calculating with the data found in table 14, we find that X is equal to 19.15km of length capable of receiving rubberised asphalt technology with the application of the waste collected in the municipality of Natal every month. The 15,000 is the number of waste materials collected each day times the number of days in the month. Also, quantifying the asphalt mix required for the length found previously equal to 19.15km and using the data cited in Table 10 as an example, let's look at the data in Table 15 below.

Table 15 - Asphalt mix for 19.15km length

Surfacing - Rubberised Asphalt
$19{,}150m \times 9.0m \times 0.035m \times 2.425Vm^3 = 14{,}628.211$
Rubberised Asphalt Putty
14,628.21 tonnes

Source: Adapted from SOUZA (2016).

Table 15 shows that 14,628.21 tonnes of asphalt are produced from the tyres collected each month in Natal/RN.

CHAPTER 5

CONCLUSIONS

In accordance with the aim of this monograph, theoretical and practical information was gathered on the environmental impacts and economic viability of rubber asphalt technology. Based on the knowledge acquired, a case study was carried out looking at the economic and environmental viability of using scrap tyres in the asphalt paving of a highway and an avenue in Natal. A comparison was made between conventional asphalt and rubberised asphalt, establishing the environmental impacts and costs. It is therefore concluded that, in general, the objectives were achieved and demonstrated during the development of this monograph. The point of convergence of this study was the demonstration of costs through execution and maintenance data of the types of pavement compared in this work, presenting findings in economic and financial improvements in the development and innovation of technology on a motorway and on an avenue in Natal in Rio Grande do Norte.

Rubber asphalt is a safer, sustainable solution for Brazilian streets and highways. The material, which is produced by adding shredded rubber extracted from waste tyres and added to the asphalt binder, has numerous advantages such as increasing the useful life of roads. Paving with rubberised asphalt also has the advantage from an economic point of view, since it is highly durable and the project, in the long term, ends up diluting the initial investment, which is due to the savings in maintenance investment for this type of pavement. In other words, in this study we saw that conventional asphalt becomes 18.4% more expensive than rubber asphalt due to the fact that there is greater wear and tear on the CBUQ, which adds up to a greater need for maintenance when compared to rubber asphalt.

Therefore, according to the results obtained in this study, it can be concluded that the economic and environmental viability of using waste tyres for asphalt paving on a road and avenue in the city of Natal (RN) is favourable, as it is environmentally and economically viable. In addition, it is possible to carry out pavement maintenance using waste tyres collected in the municipality itself.

CHAPTER 6

REFERENCES

HOT ASPHALT. **Terrena:** asphalts and paving. Available at:
<http://terrenaasfaltos.com.br/servicos-de-pavimentacao/asfalto-quente-cbuq/>. Accessed on: 11 Nov. 2017.

BRAZILIAN ASSOCIATION OF TECHNICAL STANDARDS (ABNT). **NBR 10.004:** Solid waste - Classification: Procedure. Rio de Janeiro: ABNT, 2004.

BALBO, José Tadeu. Asphalt paving: materials, projects and restoration. Oficina de texto, 2007.

BALDO, José Tadeu. **Asphalt paving:** materials, design and restoration. São Paulo: Oficina de Textos, 2007.

BERNUCCI, L. B. et al. **Asphalt paving:** basic training for engineers. Rio de Janeiro: Abeda, 2007.

BERTOLLO, S. A. M. et al. Asphalt paving: an alternative for the reuse of used tyres. **Public Cleaning Magazine,** Brazilian Association of Public Cleaning. ABPL, n. 54, 2000.

General Coordination of Maintenance and Restoration. **CREMA Project.** Natal: Crema Programme Management Consortium Contract: Tt-268/2011-00 Tender: 223/2010-00, 2015. 304 p.

CONCER. **Works in progress.** 2009. Available at:
<www.concer.com.br/obras_andamento_asf_eco.htm>. Accessed on: 20 August 2015.

NATIONAL TRANSPORT CONFEDERATION (CNT). **Road Research.** 2017.

NATIONAL TRANSPORT CONFEDERATION (CNT). **Pesquisa CNT de rodovias 2016:** relatório gerencial. 20. ed. Brasília: CNT: SEST: SENAT, 2016.
Available at:
<http://pesquisarodoviascms.cnt.org.br/Relatorio%20Geral/Pesquisa%20CNT%20(2 016)%20-%20LOW.pdf>. Accessed on: 13 Nov. 2017.

National Environment Council (CONAMA). Resolution No. 258 of 1999.
DJAILDO. **National tyre collection mobilisation continues in RN municipalities.** 2016.
Available at: <http://djaildo.com/mobilizacao-nacional-de-coleta-de- pneus-continua-nos-municipios-do-rn/>. Accessed on: 18 September 2017.

DNIT. **Paving Manual.** 3. ed. Rio de Janeiro, 2006.

FACCIO, C. F. **Reciclanip explains how tyre collection works.** 01 07 2017.

FERREIRA, George Porto (Org.). **pneumatics report.** 2017. Available at:
<http://www.ibama.gov.br/phocadownload/pneus/relatoriopneumaticos/ibama- relatorio-
pneumaticos-2017-nov.pdf>. Accessed on: 11 Nov. 2017.

FIGUEIREDO, Matheus. Rubberised asphalt from an environmental perspective. **Sustainable
Life,**
20 Aug. 2013. Available at:
<http://sustentareviver.blogspot.com.br/2013/08/asfalto-borracha-pela-otica- ambiental.html>.
Accessed on: 11 Nov. 2017.

GOMES, J. A.; OGURA, S. K. **Bibliographical survey:** treatment and reuse of used tyres. São
Paulo: Cetesb, 1993.

GRECA ASFALTOS. **Comparative study of the performance of resurfacing using asphalt-
rubber.** 2006. Available at: <www.grecaasfaltos.com.br>. Accessed on: 12 Nov. 2017.

GRECA ASFALTOS. **Fatos Asfaltos, quarterly newsletter,** year 2, n. 5 October 2015.

IMD. Available at: <www.relatorioglobalcompetitividade2016.com>. Accessed on: 12 Nov. 2017.

BRAZILIAN INSTITUTE FOR THE ENVIRONMENT AND RENEWABLE NATURAL RESOURCES
(IBAMA). **Report 2016.** São Paulo, 2017.

LACERDA, Mariana. **Tyres that turn into asphalt.** 2016. Available at:
<https://super.abril.com.br/ideias/pneus-que-viram-asfalto/>. Accessed on: 25 June 2018.

MAZZONETTO, Caroline. Asphalt-rubber. **Urban infrastructure:** projects, costs and construction,
Dec. 2011. Available at:
<http://infraestruturaurbana17.pini.com.br/solucoes-tecnicas/11/asfalto-borracha-a- adicao-de-po-
de-borracha-extraido-de-245173-1.aspx>. Accessed on: 29 August 2017.

MESQUITA, Thiago. **City Hall collects and correctly disposes of 1,100 tonnes of waste per
day in Natal.** 2016. 12 f. Natal City Hall, Natal, 2016. Available at:
<http://natal.rn.gov.br/noticia/ntc-24750.html>. Accessed on: 13 Nov. 2017.

MORILHA, Júnior. **GRECA ASFALTOS' ASPHALT-RUBBER.** 2016. Available at:
<http://www.grecaasfaltos.com.br/blog/a-historia-do-ecoflex/>. Accessed on: 18 September 2017.

NAKAMURA, Juliana. **The types of surfacing, the machinery required and the precautions to be taken when contracting, designing and carrying out.** 2011. Available at: <http://infraestruturaurbana17.pini.com.br/solucoes-tecnicas/16/pavimentacao- asfaltica-os-tipos-de-revestimentos-o-maquinario-necessario-260588-1.aspx>. Accessed on: 29 September 2017.

NÓBREGA, Eduardo Suassuna. **Comparison of retroanalysis methods for asphalt pavements.** 2014. 384 f. Thesis (Doctorate) - Civil Engineering Course, Postgraduate Programme, Federal University of Rio de Janeiro, Rio de Janeiro, 2014.

ODA, Sandra; FERNANDES JÚNIOR, José Leomar. Tyre rubber as a modifier for asphalt cements for use in paving works. **Acta Scientiarum,** Maringá, v. 23, n. 6, p. 1589-1599, 2001.

ORMOND, J. G. P. Glossary of terms used in agriculture, forestry and environmental sciences. Rio de Janeiro: BNDES, 2004.

PENA, Rodolfo F. Alves. Sustainable development. **Brazil School.** Available at <http://brasilescola.uol.com.br/geografia/desenvolvimento-sustentavel.htm>. Accessed on: 17 Nov. 2017.

PROF. CARLOS ARRUDA (Brazil). Dom Cabral Foundation. **BRAZIL NEEDS TO CREATE A POSITIVE AGENDA TO REVERSE THE DOWNWARD TREND IN GLOBAL COMPETITIVENESS RANKING.** 2016. Available at: <http://acervo.ci.fdc.org.br/AcervoDigital/Relat0rios Research/Research Reports 2017/27 05 2017 - Executive Summary - IMD Ranking Without Embargo.pdf>. Accessed on: 22 September 2017.

GROWTH ACCELERATION PROGRAMME (PAC). **Growth Acceleration Programme Report 2016.** São Paulo, 2017.

MOTORWAYS AND ROADS MAGAZINE. Asphalt Technology. Issue 36. Available at: http://www2.rodoviasevias.com.br/revista/materias.php?id=268&edicao=Edicao36; accessed on 28 May 2012.

Municipal Department of Public Works and Infrastructure (SEMOV).

SENÇO, Wlastermiler de. **Manual de técnicas de pavimentação,** v. 1.2. ed. São Paulo: Pini, 2001.

Construction Reference Cost System (SICRO).

SPECHT, Luciano. **Environmental and economic advantages of using rubber in asphalt.2016.** Available at: <http://inovacao.sielo.br/pdf/inov/v3n3/a08v3n3.pdf>. Accessed on: 22 September

2017.

URBANA collected 100,000 tyres in Natal in 2010. **Tribuna do Norte,** 1 Dec. 2011. Available at: <http://www.tribunadonorte.com.br/noticia/urbana-recolhe-100- mil-pneus-em-natal-no-ano-2010/169931 > Accessed on: 11 Nov. 2017.

WICKBOLDT, V. S. **Accelerated pavement tests to evaluate the performance of asphalt resurfacing.** 2005. 134 f. Dissertation (Master's Degree) - Postgraduate Programme in Civil Engineering, Federal University of Rio Grande do Sul, Porto Alegre, 2005.

yes **I want** morebooks!

Buy your books fast and straightforward online - at one of world's fastest growing online book stores! Environmentally sound due to Print-on-Demand technologies.

Buy your books online at
www.morebooks.shop

Kaufen Sie Ihre Bücher schnell und unkompliziert online – auf einer der am schnellsten wachsenden Buchhandelsplattformen weltweit! Dank Print-On-Demand umwelt- und ressourcenschonend produzi ert.

Bücher schneller online kaufen
www.morebooks.shop

Printed by Books on Demand GmbH, Norderstedt / Germany